Penguins

Barbara Todd

REED

CONTENTS

INTRODUCTION

'There are only two kinds of penguin: the white ones walking towards you and the black ones walking away from you!' These were the words of one Antarctic explorer who, along with other early visitors to the ice, sometimes encountered thousands of penguins.

Penguins created an interesting diversion for these early travellers, and they continue to both amuse and entertain humans. Their 'formal' dress and waddling, upright gait remind us of funny-looking humans; in fact, eighteenth-century explorer Sir Francis Drake announced, 'They walk so upright, a far-off man would take them to be awkward little children.' This 'awkward' walk has been copied by comedians and clowns, and charming images of penguins often appear in cartoons and books, on television advertisements and as logos for businesses.

While early explorers may have found penguins entertaining, they primarily viewed them as an easy source of food to supplement their diets. Penguin breast became 'palatable when correctly prepared' and eggs were turned into 'large and satisfying omelets'. Indigenous tribes, sealers and whalers also used penguins and their eggs as a food source. Sealers and whalers had the greatest impact on penguin populations, slaughtering thousands of

Did you know?

- The very first 'penguins' encountered by Europeans were not penguins at all! Explorers sailing to Newfoundland in 1578 stopped for fresh provisions at an island called Penguin. There were numerous birds that they called 'penguins' on the island. These large flightless birds were not the penguins we know today: they were Great Auks. Unfortunately, these fat and clumsy birds proved easy prey for hungry sailors and they became extinct in 1844. It is thought the birds that we now know as penguins received their name from the seamen who later ventured into the southern hemisphere and encountered large, fat, black and white flightless birds. These birds resembled the Great Auk and were also called penguins.

Great Auk

birds during the 1800s for the oil that could be extracted from their blubber. The mass killing ended just in time for species such as the Royal penguin on Macquarie Island, which was on the road to extinction.

Today, scientists study penguins throughout their southern hemisphere range. The results of their work have given us new knowledge and insight into the world of the bird that comes 'already dressed for dinner'.

When early Antarctic explorers built their huts close to or among penguin colonies, both the birds and their eggs became an easily accessible source of food.

Penguins from the Past

Some of the ancient penguins were seriously big! The largest penguin fossil was found in Antarctica. Scientists believe the penguin may have stood 170 cm tall — about the height of a grown man. This extinct penguin was given the name *Anthropornis* which means 'the man bird'. Another large fossil was discovered in New Zealand. *Pachydyptes ponderosus* may have reached a height of 164 cm. These giant penguins from the past would have towered over today's tallest penguin, the 110–130 cm Emperor penguin, and dwarfed the tiny 40–45 cm Blue penguin.

Scientists have discovered about 35 fossil penguin species. While many of the ancient penguins were larger than the penguins of today, some of them were as tiny as the Blue penguin. Scientists believe penguins evolved over 60 million years ago, from the ancestors of albatrosses, petrels and shearwaters. They appear to be most closely related to small diving petrels. Blue penguin chicks have the same tubular nostrils as albatrosses and petrels.

Although they have lost the ability to fly in the air, penguins are capable of some amazing 'aquabatics', often appearing to fly underwater. Their wings have evolved into paddle-like flippers, which are very efficient for swimming and manoeuvring their torpedo-shaped bodies through the water. Penguins use their tail and feet as a rudder and will sometimes 'porpoise' or leap clear of the sea when they are swimming quickly. Some early sea voyagers called these birds 'feathered flying fish'.

Taroa and Tawaki

Maori have a legend about the albatross and the penguin. It seemed that Taroa, the albatross, and Tawaki, the penguin, were always fighting over who was better at flying and fishing. Their bickering continued for many years until Tane Mahuta, god of the forests and birds, settled the matter once and for all. Tane Mahuta gave Taroa the gift of the longest wings of any seabird. Now Taroa could sail on the ocean winds far from land in his search for food. Tawaki received flipper-like wings and, although the penguin could no longer fly in the air, it could use its 'wings' to fly beneath the ocean waves when chasing and catching small fish.

Emperor penguin *Anthropornis* Blue penguin *Pachydyptes ponderosus*

Modern-day penguins

Scientists currently recognise eighteen types of penguin. These 'modern' penguins are scattered throughout the oceans of the southern hemisphere. Only two penguin species breed on the main Antarctic continent, while others have rookeries on the Antarctic Peninsula or islands off the Antarctic coast. Some penguin species nest in the subantarctic or in more temperate or tropical regions.

Cool and not so cool penguins

The two 'coolest' penguins, the Emperor and Adelie, mate and rear their chicks on the main Antarctic continent, Antarctic Peninsula and islands off the coast. Other 'cool' penguins, the Chinstrap and Gentoo, often nest alongside the Adelie, on the peninsula or on nearby islands. King, Royal and Macaroni penguins are also 'cool' and nest on islands off Antarctica and in the subantarctic.

'Pretty cool' penguins, like the Rockhopper, Snares crested and Erect-crested penguins, prefer subantarctic islands. The Magellanic penguin is also 'pretty cool' and is found around South America and on the Falkland Islands.

'Moderately cool' penguins, like the Yellow-eyed and Fiordland crested penguins, nest on New Zealand's South and Stewart islands. Yellow-eyed penguins also nest on the Campbell and Auckland islands (in the subantarctic). The Blue penguin lives in both New Zealand and Australia.

Emperor penguins Kim Westerskov/Hedgehog House

Some like it hot!

Cape penguins sunbathe off the southern coast of Africa, while the Humboldt penguin prefers Peru and northern Chile on the west coast of South America. The 'hottest' of them all, the Galapagos penguin, lives near the equator where temperatures may reach 40°C and unprotected eggs can become hardboiled!

How do I look?

Penguins come ready attired in a feathered black and white coat. This formal dress is not designed so penguins look sophisticated, but is actually a form of camouflage. When a penguin is swimming, its black back blends with the darkness of the ocean while its white underparts are hard to see against a background of sunlit water. The camouflage, known as countershading, gives the illusion of a grey shadow and helps to hide the penguin from both its predators and its prey. All penguins have a black and white body, but each species has distinctive and often coloured plumage around its head, and sometimes on its neck and chest.

A penguin's plump body is insulated with a layer of blubber. Penguins have a downy underlay of feathers next to their skin that helps trap body heat in cool and cold climates. The feathers in the top layer are stiff and small and overlap each other to act like a feathered waterproof wetsuit that keeps the bird both warm and dry. Flippers are covered with short, scale-like feathers and there are no feathers on the beak or feet. Penguins living in warm climates can become too hot and must lose some of their body heat through their flippers, beaks and feet. Penguins from colder areas have shorter flippers and beaks and extra facial feathers in order to retain more of their body heat.

This scruffy looking penguin is moulting. All penguins lose and replace their feathers once a year. The moult takes place over a period of weeks and the penguin must stay on land, without food, until its new coat grows back. Moulting birds are very vulnerable and will lose up to 50 percent of their body weight before they are able to return to the sea.

LIVING TOGETHER

A. Cox, DoC

Erect-crested penguins

Imagine a 'sea' of penguins on land! Some colonies host extraordinarily large numbers of penguins that breed and rear their chicks in close proximity to one another. For example, about 600,000 King penguins breed on the Crozet Islands and approximately five million Chinstrap penguins live around the South Sandwich Islands. It can get very crowded! If you could speed up the motion of a large penguin colony heading out for breakfast, it would probably resemble a crowded motorway in the morning. Nearly two million Royal penguins live on Macquarie Island. Imagine having to make your way through 1,999,999 other penguins in order to reach the sea — it makes getting off a crowded bus seem easy!

A penguin encounters another problem when it returns home with dinner for its young chick. The parent must locate its baby amongst the thousands of other hungry chicks who are all crying out, 'Feed me, feed me.' How do they do it? Penguins recognise each other by sound, not sight. Every penguin has a unique voice so a parent finds its own chick, as well as its mate, by listening for their distinctive calls. Think how confusing it must be in a penguin colony — and how noisy!

Some penguins prefer a more solitary, rural lifestyle. They nest in deep undergrowth or burrows, under rocks, inside rock caves, amongst tree roots and inside hollow logs. Yellow-eyed penguins are the least social penguin, usually nesting out of sight of one another.

What's for dinner?

Penguins don't have to 'eat their vegetables'; instead they enjoy a healthy diet of seafood that comes ready salted! They have a special gland at the base of their bill that eliminates excess salt from swallowed seawater. Penguins eat a variety of fish such as small cod, sprats, pilchards and anchovies. They also enjoy small squid and tiny shrimp-like crustaceans known as krill. Not surprisingly, the larger penguins are the deepest divers, capable of descending to depths of 350 m when they are feeding. Smaller penguins search for food that is closer to the surface of the sea.

Species preservation is the motivation for both crowded and solitary lifestyles. 'Safety in numbers' is the name of the game in crowded nesting areas. If your nest site is in the middle of the colony or is surrounded by other nests, your chick has a better chance of not being targeted during an attack from a hungry predator such as a leopard seal or skua. If your nest is hidden under bushes or rocks, your chick has less chance of being detected by stoats or other predatory animals.

Lou Sanson

Snares crested penguins

DATING, MATING AND PARENTHOOD

All penguins exhibit some form of courtship display. Some males pump out their chests in order to look more manly while other penguins stretch their wings, bow to each other and then burst into loud song, a form of penguin serenade. Crested penguins flaunt their yellow head plumes by shaking their heads from side to side.

Yellow-eyed penguins

King penguins have a comical and drawn-out courtship. They first engage in mutual sky-pointing with their flippers and then burst into an out-of-tune duet. These preliminary displays are followed by the 'penguin waltz', where the male and female perform an elaborate 'waddle' through the colony. After thanking each other for the dance with a formal bow, the penguins proceed to mate. It is not always harmonious on the mating grounds. Squabbles will sometimes erupt into fully-fledged fights, accompanied by loud squawking and flipper slapping.

Proud parents

Parenthood begins as soon as an egg is laid. The egg must be protected from the elements as well as from predators. Normally, one parent will guard the egg or eggs, while the other goes to sea to forage.

Emperor and King penguins lay one egg that is incubated on top of their feet. The egg is covered and kept warm by the penguin's thick belly skin. It takes approximately two months for the egg to hatch and the new chick continues to live on its parents' feet for four to six weeks.

Lou Sanson

Snares crested penguins

All other penguin species lay two eggs. Some penguins construct elaborate nests from pebbles — one Gentoo penguin used over 1700 stones to build its nest! Other penguin nests are simpler structures made from stones, twigs and grass. Incubation ranges from 33 days for Blue penguin eggs, to up to 65 days for Emperor penguin eggs. Most penguins that lay two eggs hatch or successfully rear only one chick. Yellow-eyed penguins are an exception and have the highest success in fledging both chicks.

Young chicks are often attacked by predators and must be guarded by their parents during their first weeks of life.

Rockhopper penguin with eggs

G. Taylor, DoC

New Zealand penguins

It seems that penguins have always liked New Zealand! At least thirteen different fossil penguins have been discovered here. Seven types of penguin currently breed around the two largest islands and on smaller islands to the south.

Yellow-eyed, White-flippered, Fiordland crested, Snares crested and Erect-crested penguins are found only in New Zealand. The Blue penguin lives in both New Zealand and Australia, while Rockhopper penguins are found around subantarctic islands throughout the southern hemisphere.

New Zealand and the subantarctic islands — penguin distribution

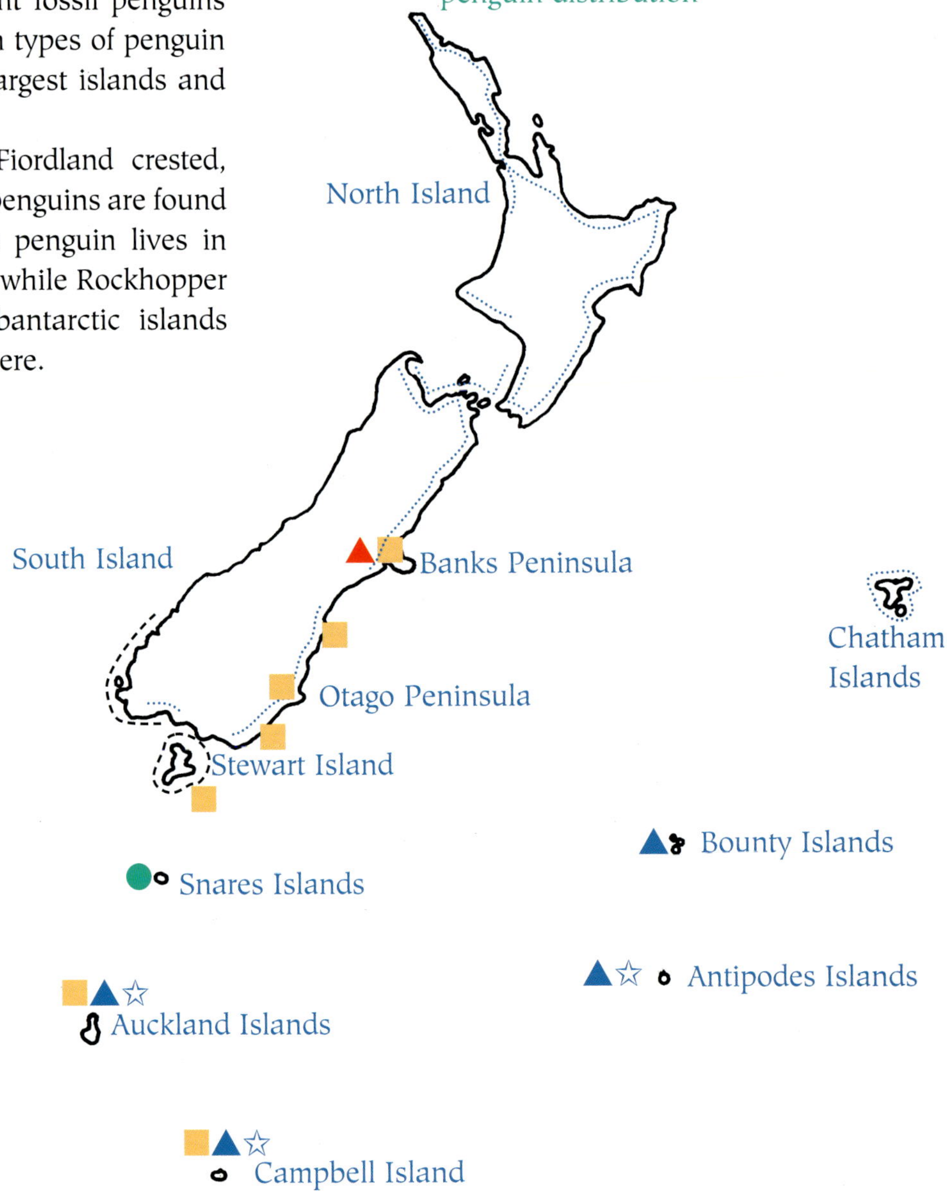

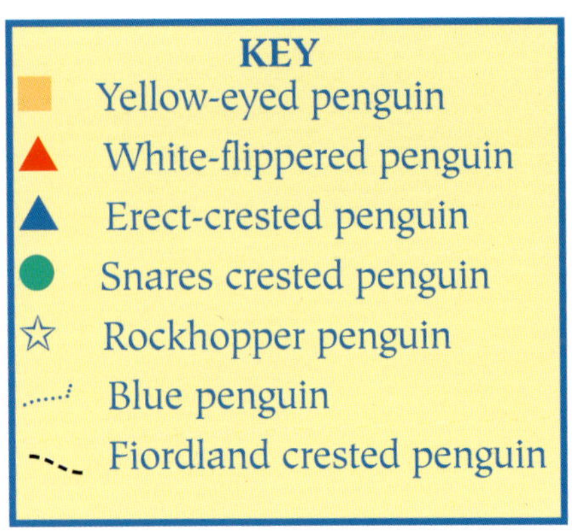

Yellow-eyed penguins

Megadyptes antipodes — 'large southern diver'
Length 65–75 cm
Weight 5–8 kg
Eggs Two, laid mid-September to mid-October
Incubation 38–52 days
Fledge Late February to early March

Yellow-eyed penguins are the third tallest penguin species. Maori call them hoiho, which means noise-shouter. It's a good name as Yellow-eyed penguins are seriously noisy! In the evening, when they return home from the sea, and again at dawn, they communicate with each other using loud and very shrill voices. Earplugs would be a good idea if a hoiho family moved next door to you!

The penguins make a daily trek from their nests to the sea to forage for food. They sometimes dive as deep as 100 m in search of squid and small fish such as sprat, silversides and red and blue cod. They are homebodies who return to their nest each evening throughout the year. Nesting sites are located along the southeast coast of the South Island, on Stewart and other small offshore islands and in the subantarctic on both Campbell and Auckland islands.

Yellow-eyed penguins like their privacy. They build isolated nest sites that are hidden amongst shrubs or tussock, in underground burrows, beneath boulders, or in tree hollows.

Many hoiho travel long distances before reaching home, where they engage in shrill

Peter Moore, DoC

Penguin chicks don't look anything like their parents. They are covered with fluffy down feathers, which are not waterproof. The chicks cannot fledge and go to sea until their adult plumage appears.

Not so cool facts

- Only about 15 percent of fledged chicks survive their first year at sea and return to their homes.
- Yellow-eyed penguins are one of the rarest penguin species. Threats to their future survival include introduced predators such as cats, stoats, rats, and ferrets; habitat destruction caused by humans and livestock resulting in the loss of protected nesting sites; and warming seas that result in less fish.

conversation. After a night's sleep, they awaken with more piercing calls and begin their morning trek back to the sea.

Yellow-eyed penguins start 'dating' when the females are 2–3 years old and the males are 3–5 years old. Serious courtship begins in August.

Once eggs are laid both parents share incubation duties. The two chicks start hatching within hours of each other and it often takes them all day to break totally free from their shell.

The new parents take turns guarding their young for six weeks; one parent 'chicksits' while the other goes out to sea 'grocery shopping'. The chicks are fed with yummy, half-digested fish, which is regurgitated into their mouths. After six weeks the chicks are left on their own while both parents go fishing in order to satisfy their babies' greedy appetites.

The chicks grow quickly on their rich diet of 'seafood soup', and weigh almost 5.5 kg when they reach their fledgling age of 4–5 months.

When they enter the sea for the first time, it must feel very cold — and very lonely! As the new chicks mature during their first year, they can swim up to 500 km north of their fledging site and will not return to their home breeding areas until the start of their second year.

Cool and interesting Yellow-eyed penguin facts

Yellow-eyed penguins:

- are found only in New Zealand,
- may be one of the most ancient of all living penguins,
- sometimes make a daily round trip of 30 km searching for food,
- dive 200–300 times a day while foraging for prey,
- usually mate for life ... although divorce is not unheard of,
- lose up to half of their body weight (3–4 kg) when they moult.

Yellow-eyed penguins are the only penguins with yellow eyes.

'Hurry up, honey ... ' This penguin pair has a long trek ahead and will scramble up muddy slopes and rocky outcrops before reaching their nest, which is almost a kilometre from the sea.

Blue penguins — the 'fairy' penguins

Eudyptula minor — 'good little diver'
Length 40–45 cm
Weight 1.1–1.5 kg
Eggs Two, laid September to October
Incubation 33–40 days
Fledge November to December

Blue penguins are also known as Little Blue or Fairy penguins. 'Little' and 'fairy' say it all, for these penguins are very tiny. Their Maori name is koura.

Blue penguins will dive to depths of 30 m searching for their primary prey of small fish. They are opportunistic feeders, and will add squid or krill to their diet in areas where it is available. Blue penguins live in both New Zealand and Australia. The name 'Fairy' penguin is most commonly used in Australia.

The Blue penguin is not such a good little fairy if it happens to breed close to where you are living. Although they often nest in ground burrows or under flax bushes, tussocks or rocks, they have been known to select cosy dark sites under houses. The male selects the nest site and then stands at its entrance singing the 'penguin love call'. The call consists of braying, barking, yelping and growling — it may sound lovely to a female penguin, but it's not very nice if you're a human. Once a female is attracted, the noise continues while the pair mutually serenade each other to formalise their bond. When a pair has bonded they usually return to the same nesting site every year.

Blue penguins lay two, tiny 53-g eggs. (Check out 50 g of butter and you'll feel how much each egg weighs.) The parents take turns incubating the eggs. The chicks are born blind and almost naked and must be constantly guarded by one of the parents for about three weeks. The young chicks are then left on their own and wait impatiently for their nightly ration of food. Sibling rivalry occurs as the chicks stand at the entrance to their home, squabbling and competing for the first and biggest dinner. The largest chick is sometimes the only one who survives to adulthood.

After eight weeks, the fat and fully feathered chicks are ready to fledge. The parents then spend a few weeks at sea gorging themselves in preparation for their two- to three-week yearly moult.

Right: The White-flippered penguin causes some confusion for scientists. Some consider it a separate species, while others claim it is a subspecies of the Blue penguin. The White-flippered penguin is a little heavier and has a conspicuous white band around its flippers. It lives in colonies on Banks Peninsula and at adjacent areas around the east coast of the South Island.

R. Morris, DoC

T. Lilleby, DoC

Cool and interesting Blue penguin facts

Blue penguins:

- are the smallest penguin species,
- have been recorded diving to depths of 69 m,
- consume approximately 90 kg of food every year,
- sometimes feed co-operatively. They work together to herd shoals of small fish into a shallow bay where they dive through the middle of the shoal and gobble up the fish.

Crested penguins — the 'punk' penguins

Eudyptes — 'good diver'
Length 55–65 cm
Weight 3–6 kg
Eggs Two, laid August to November
Incubation 33–36 days
Fledge Late November to mid-January

Crested penguins are very fashionable. Their bright yellow tufted 'eyebrows' give them a distinctive and jaunty appearance. Each type of crested penguin has its own style of head plume that they display and flaunt. They will often shake their tassels excitedly, as if to say, 'Hey, check me out', when they are attempting to attract a mate.

Brent McFadden

Cool and interesting crested penguin facts

- The male Snares crested penguin fasts for five weeks while on parental duty, incubating eggs and then guarding new chicks.
- Snares crested penguins sometimes build their nests in the low branches of trees — the penguin equivalent of a high-rise apartment?
- Male Erect-crested penguins are the home-builders. They often pilfer material from neighbouring nests and add it to their own dwelling of pebbles, twigs, leaves and the odd bird feather.
- Rockhopper penguins are mini-mountaineers and 'hop' up and down rocks that are as tall as they are.
- Courting Rockhopper penguins hold the record for the fastest crested penguin headshakes.

Crested penguins are medium-sized birds who nest in loose colonies amongst bushes and tree roots, in rock caves or under boulders. They lay two eggs but, unlike other species that lay two eggs, their second egg is larger than the first. Although two chicks are usually hatched, the second and larger chick is normally the only one to survive. The date of egg-laying varies for each species and depends on where they live.

A. Munn, DoC

Penguins in warmer climates lay their eggs earlier in the year than those in colder areas.

Parents take turns guarding new chicks and going to sea to forage for food. At approximately 21 days, the chicks resemble fat, furry balls and they are left in a crèche with other 'furry balls' while both parents go to sea. The prey of crested penguins varies. They feed on small fish, squid or krill, depending on the area in which they live.

There are six species of crested penguin. Two species, Macaroni and Royal penguins, live outside New Zealand. Rockhopper penguins nest both within and outside New Zealand. Fiordland crested, Erect-crested and Snares crested penguins live and breed only in New Zealand.

Fiordland crested penguins

Fiordland crested penguins are found around the southwest coast of the South Island and on Stewart and Solander islands. Like all crested penguins, they have attractive yellow head plumes. One of their distinguishing features is a small patch of white cheek feathers. They are not as sociable as some of the other crested species and build their nests about two metres apart. When they are mating the birds engage in mutual displays and issue a series of trumpeting calls while pointing their bill skyward and shaking their heads from side to side. When their chicks hatch, the male is a 'house dad' and guards them for the first 21 days. Chicks fledge in late November when they are 75 to 80 days old. The parents, like all other penguin species, must then build up their fat reserves in anticipation of their moult, which begins in February and lasts until March.

Lou Sanson

Snares crested penguins

Snares crested penguins are found mainly on New Zealand's Snares Islands, in the subantarctic. They are similar in appearance to the Fiordland crested penguin, but are often heavier, weighing up to 5 kg. The robust size of their bill gave them the name *Eudyptes robustus*. Their yellow crest starts at the bill's base and then forms eyebrows before fanning out at the back of the head. Like all crested penguins, they flaunt their head plumes during courtship. The males shake their heads up and down and emit loud calls while standing with wide-open flippers, as if to say, 'Come to me, baby.' Once the female is attracted, the pair engage in pulsating vocalisations and mutual displays. Snares crested penguins build large and roomy nests among the islands' inland forests. The nests are often lined with leaves for additional comfort. Eggs are laid from late September to mid-October and incubation is time-shared.

The male takes the final thirteen-day shift and then guards the new chicks for another three weeks. The chicks leave the colony when they are around 75 days old, followed by their parents who must fatten up before beginning their April moult.

Erect-crested penguins

Erect-crested penguins have large colonies on the Bounty and Antipodes islands, and smaller colonies on both Campbell and Auckland islands. The plumes of the Erect-crested penguin are upturned and form a brush-like crest. They have a noisy braying call, which they use

R. Morris, DoC

J. Kendrick, DoC

excessively during courtship. Erect-crested penguins don't seem to mind company and will often build their nests close together. The parents take turns incubating the two eggs that are usually laid during October. After the chicks are born, the parents continue to share guarding and feeding duties. Like other crested chicks, the young quarrel noisily in their crèches while waiting for their regurgitated 'food packs'. The 45–50-day-old chicks head to sea in mid-January, followed by their parents who return to the breeding site in March to begin their month-long moult. By late April the colony is deserted, until the next breeding season.

Rockhopper penguins

Rockhopper penguins are the smallest crested penguin, but one of the most aggressive. They have rather untidy, long, drooping plumes, a spiky crest of black feathers on top of their head and red eyes ... too many late nights perhaps?

Rockhopper penguins live up to their name, hopping up rocky inclines and jumping off low rocks. They look quite strange jumping down rocky slopes with their feet held closely together, but they move with seeming ease and efficiency. Some of their colonies are located on the top of rocky cliffs.

Rockhopper penguins are the most plentiful crested penguin and have undergone 'urban sprawl'. Their large, widely distributed breeding colonies are located on islands throughout the subantarctic. In New Zealand they are found mainly on Campbell Island, with additional smaller colonies located on the Antipodes and Auckland islands. Eggs are laid in early November and parents share incubation duties. The dedicated father 'chicksits' the newly hatched young for three weeks and Mum takes over the first week of crèche feeding. Chicks go to sea when they are about 65–70 days old. The parents complete their moult by June, and spend the winter months foraging at sea. Rockhopper penguin colonies have experienced dramatic losses over the last 30 years, possibly as a result of food shortages.

G. Taylor, DoC

OTHER PENGUINS

There are thirteen types of penguin that breed outside of New Zealand: this number includes the Blue and Rockhopper penguins that also live and breed in New Zealand waters. Their territories range from the Antarctic to the tropics and include nesting sites on subantarctic islands throughout the southern hemisphere.

Two species of crested penguin, the Macaroni and the Royal, breed exclusively outside New Zealand. Macaroni penguins are one of the most gregarious of the crested penguins. They were named after a group of trendy male fashion-setters who wore very high wigs and tiny hats. These 'fashionable' young men formed the London Macaroni Club in the late 1700s. When a penguin was discovered which had flamboyant bright yellow-orange crests, it seemed natural to call it the Macaroni penguin. Macaroni are one of the most abundant penguins, with an estimated twelve million pairs breeding on sub-antarctic islands in the South Atlantic and Indian Ocean. Royal penguins are also flashy lookers and resemble the Macaroni; in fact, some scientists consider them a subspecies. Royal rookeries are on the Australian sub-antarctic island of Macquarie. Royals were hunted extensively in the early 1900s, when up to 150,000 a year were killed for their oil. Their population has now recovered and is estimated at approximately two million.

Emperor penguins — the extreme penguins

Aptenodytes — 'flightless diver'
Length 110–130 cm
Weight 30–40 kg
Eggs One, laid in May
Incubation 62–65 days
Fledge Late November to mid-December

Emperors are called the 'extreme penguin' for a variety of reasons. They are the tallest, the heaviest, the deepest undersea penguin diver and live in one of the coldest places on earth. Their breeding sites, on the Antarctic continent and surrounding islands, are so remote they weren't discovered until Scott's first expedition to Antarctica, in 1903.

Between March and April, Emperor penguins trudge in long processions across the ice pack to gather at their breeding sites. Their courtship is a formal and courteous one, comprising synchronised calls and the 'penguin waddle dance', followed by bowing and mating. The calls are important as they enable penguins to recognise each other when they return to the colony after foraging at sea.

There is no place to build a nest and no material to build one with on the ice. However, it doesn't matter because Emperors have a built-in nest — their feet. The female lays a single 450-g egg which is transferred to the top of the male's feet and tucked under a rich, warm belly-fold. The egg is incubated during the harsh Antarctic winter. The males huddle together for warmth in temperatures as low as – 62°C and winds that reach 180 km/h!

When the nearly naked chick hatches, the male feeds it with a milky substance and tucks it back under its warm belly. The male has fasted for up to 110 days by the time the female returns with food for the chick. The hungry male transfers the new offspring to his mate's feet and goes to sea for a well-earned feed. He returns in about three weeks and another chick transfer occurs. After 40 days, the chick is left huddling in a crèche while both parents forage for its food.

Chicks leave the colony when they are 120–150 days old. It is early summer and the waters of Antarctica are rich with food. Both chicks and adults gorge themselves, the adults anticipating their yearly moult that begins in December and lasts for 30–40 days.

King penguins — the most photogenic of all?

Aptenodytes — 'flightless diver'
Length 85–95 cm
Weight 12–16 kg
Eggs One, laid November to March, twice every three years
Incubation 56 days

Lou Sanson

King penguins share the same genus as Emperors and are sometimes confused with them. The King penguin is smaller, but, as befits a 'king', its colours are brighter and more flamboyant. They breed on subantarctic islands around the Southern Ocean and, like the Emperor, they incubate their single egg and shelter the new chick on their feet.

The King penguin's breeding cycle is unlike any other species'. They mate only twice every three years. After laying a 300-g egg, the female places it on her mate's feet and goes to sea for about 20 days. After her first long period at sea, both parents share incubation. It is a nervous time, as the egg must be transferred safely from the feet of one parent to the other — it doesn't always work!

The nearly naked new chick is soon covered with a layer of brown down. It remains on its parents' feet for 3–5 weeks. The chicks then go into large crèches where they are fed frequently. At two months of age, the chick is grotesquely fat and almost as large as its parents. During the winter months, the parents return only four or five times. The young chicks huddle together to protect themselves from violent subantarctic storms and predators such as skua and giant petrels. Only the strongest survive; up to half of the chicks die during this period. The parents return more frequently around October, when the weather becomes milder. The surviving chicks gain weight rapidly and will finally fledge in mid- to late-November.

Brush-tailed penguins

Pygoscelis — 'elbow leg'
Length 70–76 cm
Weight 4.5–6.7 kg
Eggs Two
Incubation 34–37 days

Adelie and Chinstrap penguins look similar but the Adelie penguin has a white ring around its eye while the Chinstrap has a black line running under its chin to a point behind its eye.

Adelie, Chinstrap and Gentoo penguins have long tail feathers that stick out like a brush. These 'brush-tailed' penguins are very good housekeepers, using their tails to sweep away melted snow from around their nests. They live on the Antarctic Peninsula and islands surrounding Antarctica. Only Adelie penguins breed on the main continent, and some Gentoo penguins breed on more temperate subantarctic islands. Their main diet is krill, although Gentoo penguins in warmer areas add fish to their diet.

Although brush-tailed penguins lay two eggs, only the first and strongest chick normally survives to fledgling stage.

Adelie penguin chicks fledge when they are 50–55 days old, often heading out to the open sea on melting ice floes and not returning to their nesting sites for 2–3 years. Chinstrap penguins lay their eggs from November to early December and incubation takes about 37 days. Some chicks are fed twice daily during the first few weeks, making their growth rate faster than that of Adelie or Gentoo penguins. The chicks fledge when they are between 35 and 45 days old.

Chinstrap penguins are gregarious (meaning seriously friendly!) and they nest in large colonies, often sharing their space with Adelie and Gentoo penguins, as well as Imperial cormorants. They normally reunite with the same mate every season, with pairs sometimes remaining together throughout the year.

Gentoo penguins are slightly larger than other brush-tailed penguins, their bright orange beaks giving them a more striking appearance. They are somewhat sedentary and will remain close to their nesting areas year round, especially in the warmer subantarctic regions. Their eggs are incubated for 35 days and the new chicks are guarded for 3–4 weeks before joining a crèche. Gentoo penguin chicks are slow-growing and do not fledge until they are aound 80 days old. Usually, the first and strongest chick is the only one to survive.

'Ringed' penguins

Spheniscus — 'wedge-shaped'
(refers to triangular-shaped flippers)
Eggs Two
Incubation 38–45 days

The 'ringed' penguins, which include the Magellanic, Humboldt, Cape and Galapagos penguins, live in more temperate, and sometimes even tropical climates than other penguin species. For this reason their faces are almost free of feathers, enabling them to lose more of their body heat.

Ringed penguins nest in shady areas such as burrows or rock caves, to protect their chicks from both the sun and predators. Since they are reasonably well protected, most chicks do not join crèches but remain alone on their nests while their parents are foraging for their food. The main prey of ringed penguins is small shoaling fish, such as sardines and anchovies. The Magellanic penguin is the largest ringed penguin, measuring between 68 and 73 cm and weighing between 4 and 4.8 kg. The Galapagos penguin is the smallest, with an average length of 50 cm and weighing between 2 and 2.5 kg.

Magellanic penguins breed along the coast of Argentina, around the southern tip of South America and on the Falkland Islands. One of their most notable characteristics is their call, which resembles the bray of a donkey. Chicks reared in burrows are generally left alone after the initial guard stage, while chicks in more open nest sites will join crèches. Magellanic chicks fledge in early February.

The Cape penguin is also known as the African, Black-footed or Jackass penguin. It measures between 63 and 68 cm and weighs between 3.5 and 4 kg. Like the Magellanic penguin, it has a call that resembles a braying donkey. Cape penguins are found on islands and the mainland around the southern shores of Africa and on the eastern side of Cape Horn. The parents share incubation and chick-rearing and the chicks fledge when they are about 80 days old. Due to human pressures, such as oil spills, egging and overfishing, the population of this penguin is declining.

The Humboldt or Peruvian penguin weighs in at between 4 and 4.5 kg and measures between 67 and 70 cm. Since they live in areas that have warm air temperatures and a relatively constant food supply, they can breed year round. However, when fish disappear (usually due to a rise in water temperature) their breeding success declines. Parents share chick-raising and feed their young daily, often foraging in small groups. Humboldt penguin populations are declining.

Galapagos penguins are the warmest-weather penguin. They bask in the tropics near the equator, along the west coast of Isabella and in the Galapagos Islands. Like the Humboldt penguin, their food sources and successful breeding are dependent on water temperature. To make up for their unpredictable environment, the

Cape penguins

Nathalie Patenaude

penguins will sometimes moult and breed twice a year. They have also been known to breed three times in a single year. The newly laid eggs and hatched chicks must be protected from the sun as temperatures can reach 40°C. Parents sometimes return home to hard-boiled eggs! After hatching there is intense competition between the two chicks and usually only one survives. The Galapagos is one of the rarest penguin species, with a population of around 2000 birds.

Past, present and future

The past: human exploitation

Even before the arrival of Europeans, humans hunted penguins for food on South Atlantic islands and in South America, Africa, Australia and New Zealand. Evidence from South America indicates penguins were hunted in Tierra del Fuego 6000 years ago. Early European explorers soon learned that penguin meat and eggs could be utilised to supplement their diet during their expeditions, and there are numerous accounts of shipwrecked sailors who survived on penguins and seals while awaiting rescue. While these early hunters had an impact on penguin populations, massive exploitation did not occur until the arrival of the sealers and whalers. Penguins were more than a source of food to

King penguin Lou Sanson

these hunters: they were killed for fuel and oil. A combustible source was needed for the fires used to boil down and extract oil from seal and whale blubber and fat-laden penguins soon became 'firewood', thereby conserving coal reserves. As stocks of other marine mammals dwindled, the penguins themselves were boiled down to extract their oil. These slaughters occurred throughout the southern hemisphere and some penguin populations were almost decimated.

The present

Even today, a penguin's life is not always easy. They are threatened by natural and introduced predators. Natural predators include birds such as skuas, giant petrels and gulls, who hang around the rookeries waiting for opportunistic 'penguin takeaways' — eggs and tender young chicks. Leopard seals, sea lions, sharks and orca (killer whales) also supplement their diets

A feral cat on one of the subantarctic islands. It probably escaped from a passing boat.

Leopard seal

with tasty penguin snacks. In some tropical areas, lizards and snakes prey on eggs and young chicks. These natural predators are part of nature's cycle and do not normally threaten healthy populations.

Introduced predators such as rats, stoats, ferrets and cats have a far greater impact on penguin lives. They eat both eggs and vulnerable chicks and their added presence threatens species which are already suffering from low population numbers, such as New Zealand's Yellow-eyed and Fiordland crested penguins. Humans also continue to have an impact on the lives of penguins, mainly through the destruction of their habitat. In Antarctica, roads and runways have been built across some penguin rookeries. In more settled, domestic areas, housing and roaming stock may disturb local penguin habitats. The harvesting of guano for fertiliser has reduced nesting sites for some ringed penguin species.

Additionally, modern-day penguins join other ocean inhabitants facing food shortages caused by global warming and over-fishing. Oil pollution poses a major problem and oil spills from ships and industrial spills have become all too frequent during the past 30–40 years.

Not-so-cool penguin facts

- In the late 1800s, up to 4000 King and Royal penguins were killed daily on Macquarie Island. Royal penguins nearly became extinct during this period.
- Between 1863 and 1866, an estimated 500,000 Rockhopper and Gentoo penguins were killed for their oil.
- King penguins were wiped out in the Falkland Islands by the turn of the century. A few hundred have now recolonised.
- Oil from King penguins killed on South Georgia was used to preserve fur seal skins.

There is some good news though! We are starting to wake up to the environmental problems facing both animals and humans. We are creating nature reserves for many animal species, including penguins. People have become involved with conservation projects, for example volunteers working with the Yellow-eyed Penguin Trust have planted thousands of trees and shrubs to provide protected breeding sites, while other volunteers have completed a survey of the Yellow-eyed penguin population on Stewart Island.

The future?

Environmental awareness and respect for the rights of other species are two key issues if we are to ensure their future survival. Some penguins thrive, others are in decline. It would be sad to think that any penguin species would end up like the Great Auk, which only became extinct in 1844, not really that long ago. Our responsibility to those who share our planet is to provide a healthy environment and protected habitats in which all can live and breed.

DID YOU KNOW?

- The name 'penguin' possibly came from the Spanish word 'pinguigo', which means chubby or fat.
- Penguins go tobogganing! They often travel over snowy terrain by flopping down on their bellies and pushing themselves along with their feet.
- King and Emperor penguins dive the deepest and stay under the longest. One Emperor penguin was recorded diving to 535 m, while another remained under water for eighteen minutes.
- Blue penguins only remain under water for one to two minutes.
- Foraging Gentoo penguins have been recorded making up to 90 dives in four hours. One 'Olympic' Gentoo made 460 dives in fifteen hours.
- King penguins produce up to 2 kg of new feathers when they arc moulting.
- Emperor penguins lay the largest egg at about 450 g, Blue penguin eggs are the smallest at around 50 g.
- Emperor penguins sometimes lose their egg. They have been observed attempting to incubate substitutes such as round pieces of ice, snowballs, rocks, bottles, dead fish, stale buns and even a camera case!
- Emperor penguin males incubate their egg during the Antarctic winter, when winds can reach 180 km/h and temperatures can fall to –62°C.
- Emperor penguin males return to the sea after a four-month fast, having lost up to half of their body weight.
- Adelie, Chinstrap and Gentoo penguins are petty thieves who often steal nesting sites and material from each other. One eager Gentoo bachelor penguin tried to steal nine nesting sites over a two-week period.
- Some Adelie, Chinstrap and Gentoo nests have contained pieces of tin and glass, half a bar of chocolate, the head of a teaspoon and parts of a broken mirror!